MERCURY

MARYSA STORM

BLACK RABBIT BOOKS

Bolt Jr. is published by Black Rabbit Books
P.O. Box 227, Mankato, Minnesota, 56002.
www.blackrabbitbooks.com

Michael Sellner, designer and photo researcher

Names: Storm, Marysa, author.
Title: Mercury / by Marysa Storm.
Description: Mankato, Minnesota : Black Rabbit Books, [2023]
Series: Bolt Jr. space explorer | Includes bibliographical references and index. | Audience: Ages 6-8 | Audience: Grades K-1 | Summary: "Suit up! Mercury is a planet like no other, and it's time to explore it through a colorful design that features vibrant photography, engaging infographics, and carefully leveled text"– Provided by publisher.
Identifiers: LCCN 2021022209 (print) | LCCN 2021022210 (ebook) | ISBN 9781623107758 (hardcover) | ISBN 9781644665961 (paperback) | ISBN 9781623107833 (ebook)
Subjects: LCSH: Mercury (Planet)–Juvenile literature.
Classification: LCC QB611 .S763 2023 (print) | LCC QB611 (ebook) | DDC 523.41–dc23
LC record available at https://lccn.loc.gov/2021022209
LC ebook record available at https://lccn.loc.gov/2021022210

Image Credits

123RF: 3quarks, 4–5, 12–13; chiarivfx, 10–11; forplayday, 1, 16–17; Dreamstime: Karen Foley, 12–13; iStock: buradaki, 10–11; NASA: ESA/C. Carreau, 18–19; National Air and Space Museum Collection: Transferred from NASA, Jet Propulsion Laboratory, 4–5; Science Source: Christian Darkin, 8–9; Detlev van Ravenswaay, 20–21; MARK GARLICK, 20–21; QA International, 4–5; Shutterstock: Bikomins, 3, 24; HelenField, 22–23; JLStock, 6–7; Mopic, 14–15; NASA images, 12–13; Vadim Sadovski, Cover, 6–7; Vectorfair.com, 14–15

Contents

CHAPTER 1

Meet the Planet

In 1974, a **probe** flew through space. It was the first to reach Mercury. As it passed, it took pictures. It sent them to Earth.

probe: a machine used to collect information from outer space and send it back to Earth

COMPARING SIZES

Mercury

9,525 miles (15,329 kilometers) around

Size

The pictures showed that Mercury has craters. These holes cover the planet. It looks a lot like Earth's moon. The planet is small too. It isn't much bigger than the moon.

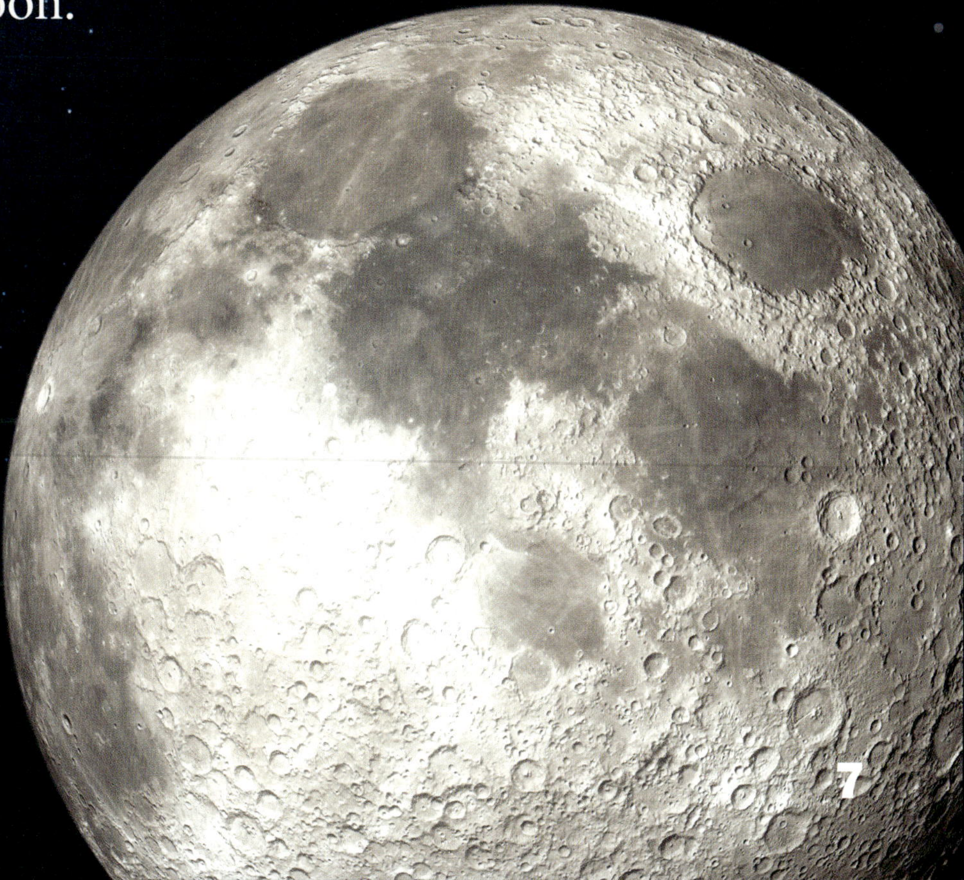

Earth's moon
6,784 miles (10,918 km) around

Features

long cliffs
rocky surface

CHAPTER 2

Its Place in Space

Mercury is the closest planet to the sun. It gets superhot. But it has a thin **atmosphere**. It can't hold in heat. Nights get very cold.

atmosphere: the gases that surround a planet

FACT

Temps drop to -279 degrees Fahrenheit (-173 degrees Celsius) at night.

Fast

This planet moves quickly. It circles the sun faster than any other. One **year** is only 88 days.

year: the time it takes a planet to make one complete trip around the sun

Mercury in the Solar System

Neptune
Uranus
Saturn
Jupiter
asteroid
belt

CHAPTER 3

Explore!

Mercury is hard to reach. It can be 138 million miles (222 million km) away. Only two spacecraft have visited. But more are on their way.

FACT

One craft crashed into the planet.

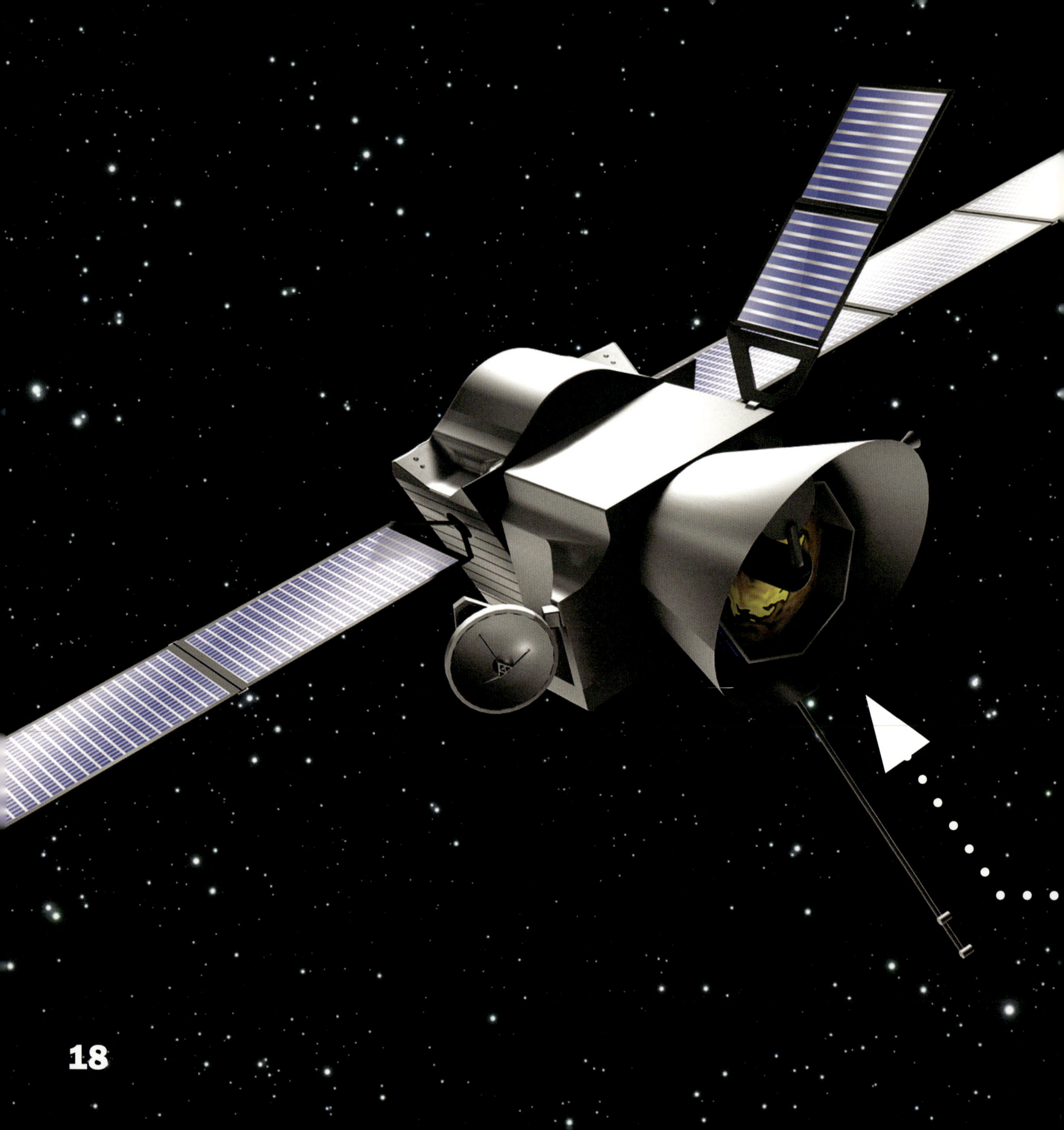

Learning More

The *BepiColombo* launched in 2018. It carries two probes. They will start circling the planet in 2025. People can't wait to see what they discover.

weight of the *BepiColombo* spacecraft

9,040 pounds (4,100 kilograms)

Bonus Facts

It doesn't have any moons.

The planet spins slowly. Its days are **1,408** hours long.

It's the solar system's smallest planet.

It has a huge **core.**

core: the central part of a planet or other body

READ MORE/WEBSITES

Boutland, Craig. *The Inner Planets.* Space Detectives. Tuscon, AZ: Brown Bear Books, 2021.

Devera, Czeena. *Mercury.* My Guide to the Planets. Ann Arbor, MI: Cherry Lake Publishing, 2020.

Grack, Rachel. *Curious about Mercury.* Curious About. Mankato, MN: Amicus/ Amicus Ink, 2022.

All about Mercury
spaceplace.nasa.gov/all-about-mercury/en/

Astronomy: Planet Mercury
www.ducksters.com/science/mercury.php

GLOSSARY

atmosphere (AT-muhs-feer)—the gases that surround a planet

core (KOHR)—the central part of a planet or other body

probe (PROHB)—a machine used to collect information from outer space and send it back to Earth

year (YEER)—the time it takes a planet to make one complete trip around the sun

INDEX